GET INVOLVED

VOLUNTEERING FOR THE ENVIRONMENT

by Chelsea Xie

BrightPoint Press

San Diego, CA

BrightPoint Press

© 2022 BrightPoint Press
an imprint of ReferencePoint Press, Inc.
Printed in the United States

For more information, contact:
BrightPoint Press
PO Box 27779
San Diego, CA 92198
www.BrightPointPress.com

ALL RIGHTS RESERVED.

No part of this work covered by the copyright hereon may be reproduced or used in any form or by any means—graphic, electronic, or mechanical, including photocopying, recording, taping, web distribution, or information storage retrieval systems—without the written permission of the publisher.

LIBRARY OF CONGRESS CATALOGING-IN-PUBLICATION DATA

Names: Xie, Chelsea, author.
Title: Volunteering for the environment / by Chelsea Xie.
Description: San Diego, CA : BrightPoint Press, [2022] | Series: Get involved | Includes bibliographical references and index. | Audience: Grades 7-9
Identifiers: LCCN 2021005748 (print) | LCCN 2021005749 (eBook) | ISBN 9781678201302 (Hardcover) | ISBN 9781678201319 (eBook)
Subjects: LCSH: Environmentalism--Juvenile literature. | Green movement--Juvenile literature. | Volunteers--Juvenile literature. | Volunteer workers in environmental protection. | Volunteer workers in government. | Volunteer workers in community development. | Volunteer workers in horticulture. | Volunteer workers in parks.
Classification: LCC GE195.5 .X54 2022 (print) | LCC GE195.5 (eBook) | DDC 333.72--dc23
LC record available at https://lccn.loc.gov/2021005748
LC eBook record available at https://lccn.loc.gov/2021005749

CONTENTS

AT A GLANCE	4
INTRODUCTION CELEBRATING ARBOR DAY	6
CHAPTER ONE HOW CAN I VOLUNTEER AT A GARDEN?	12
CHAPTER TWO HOW CAN I VOLUNTEER AT A CLEANUP?	26
CHAPTER THREE HOW CAN I VOLUNTEER AT A NATIONAL PARK?	40
CHAPTER FOUR HOW CAN I VOLUNTEER AS AN ACTIVIST?	54
Glossary	74
Source Notes	75
For Further Research	76
Related Organizations	77
Index	78
Image Credits	79
About the Author	80

AT A GLANCE

- People who are interested in garden work can volunteer at city parks and botanical gardens. They can work with others to develop a community garden.

- Volunteering at gardens includes many different tasks. Volunteers may plant flowers. They may rake leaves or pull weeds. Others may educate people about plants and the environment.

- In addition to environmental benefits, green spaces improve community life. Research shows that parks lower nearby crime rates. Gardens may encourage people to eat healthful diets.

- Organizations such as Earth Day Network and American Rivers have websites to help volunteers find cleanups. People can also organize their own cleanups.

- At cleanups, volunteers pick up waste. They sort trash from recyclables.

- National and state parks have many volunteer opportunities available. Those interested in volunteering must apply with the National Park Service or with local park authorities.

- Young people can show they care about the environment. They may write to government leaders. They can sign petitions. They may support laws that work to stop climate change.

- Starting an environmental club at school is a good way to get involved. Students can work together to tackle projects that help the environment.

INTRODUCTION

CELEBRATING ARBOR DAY

It is a warm day at the end of April. Anna is with her classmates planting trees for Arbor Day. That week during science class, Anna learned that trees are important for the environment. Trees help make the air and soil healthier. Her class also took a field trip to a tree nursery. She learned about the types of trees she sees near her home.

Arbor Day is observed in the United States on the last Friday in April. Many people plant trees on this day.

Anna had been looking forward to planting a tree all week. At last, Arbor Day has finally arrived. She and a teacher begin digging a hole for the tree. Anna takes

Volunteers sometimes plant trees at parks.

another scoop of dirt with her shovel. The hole is now big enough for a tree's root ball. The teacher helps Anna lower a young tree into the hole. They cover the roots with soil.

Anna pours water near the base of the tree. This will help it grow.

Anna wipes sweat from her forehead. Her hands are covered in dirt. Planting a tree was hard work. But Anna knows she did something good for the environment.

Her classmates soon finish planting their trees. Anna takes a step back to admire their work. She has enjoyed volunteering in the outdoors. She looks up more ways to get involved with nature when she gets back home. She wants to help the environment in any way she can.

Newly planted trees need frequent watering.

TAKING CARE OF THE ENVIRONMENT

Planting a tree is just one way to help the environment. There are many other volunteering opportunities. Some people help at gardens. Other people organize cleanups. They can volunteer at parks. Volunteering for the environment is a good way to spend time outdoors. Volunteers clean green spaces for community members to enjoy. They help make the world a healthier place for all.

CHAPTER ONE

HOW CAN I VOLUNTEER AT A GARDEN?

Gardening is a fun hobby for all ages. It is also a great way to spend time outdoors. Gardeners learn about the environment. They grow plants and take care of them. They develop green spaces for the community to enjoy.

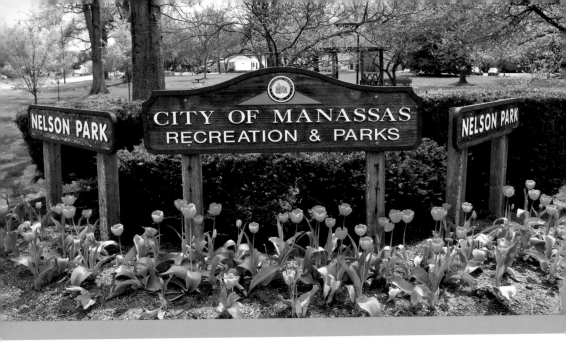

City parks may have gardens by their entrance signs. Volunteers can help keep these gardens healthy.

There are many ways to get involved at a garden. City parks have volunteering opportunities for gardeners. People can also help at botanical gardens. Others help create community gardens for easier local access to fruits and vegetables.

VOLUNTEERING AT A CITY PARK

People living in cities may not have much green space near them. For many, the easiest way to experience nature is to go to a park. Volunteering at a park allows people to develop a stronger connection with the outdoors.

City park workers welcome volunteers to help keep gardens looking fresh. Gardens take a lot of maintenance. Park employees may organize weekly events for volunteers. Volunteer tasks vary depending on the season. Often volunteers begin by picking up trash. They may work on fixing trails or weeding. Volunteers at some parks remove **invasive species** from the area.

Spring is a popular season for volunteers. Gardeners plant many colorful flowers and trees. These plants attract helpful insects such as bees and butterflies. But volunteers

COMPOSTING

Many gardens encourage composting. Compost can be used as a natural fertilizer to improve soil health. All compost is a mixture of browns, greens, and water. Browns include yard waste such as dead leaves and branches. Greens include food scraps such as vegetable waste and coffee grounds. Some materials such as black walnut leaves and meat scraps should not be used in compost. They release harmful substances and may attract pests.

are also needed in the fall. Fall volunteers help prepare the garden for the spring planting. They do tasks such as raking. They lay down mulch on garden beds.

    Park employees generally provide the necessary tools and materials. These include things like shovels and seedlings. They may have gloves for volunteers. Volunteers should wear comfortable clothing that they don't mind getting dirty. They should bring water to stay **hydrated**. It is important to put on sunscreen, since volunteers are working outdoors.

Broad-brimmed hats can protect gardeners' faces from the sun.

Volunteering at a city park has many benefits. Parks and other green spaces improve mental health. Research shows that access to green spaces lowers crime rates. Richard Sadler is a professor of **urban** geography. He said, "Places that had been greened more consistently over time saw decreases in crime over time."[1] Transforming urban areas into green spaces has a long-lasting effect on lowering crime.

VOLUNTEERING AT A BOTANICAL GARDEN

Helping at a botanical garden is another way to get involved with the environment. These gardens have a wide variety of plants from around the world. Scientists study these plants to understand how they grow. They work to protect rare types of plants.

Friends can volunteer together to help maintain a park garden.

More than 500 million people visit botanical gardens every year. They learn about the importance of plants. Humans release carbon dioxide through activities such as burning fuel. This causes climate change, which is the global increase in extreme weather and rising temperatures.

Some volunteers at botanical gardens make sure visitors stay on the paths.

Plants take in carbon dioxide and give off oxygen. This slows climate change.

Volunteers may work to educate visitors about plants. They walk around the garden and answer questions. They make sure visitors do not damage the plants.

Others work with plants directly. Their duties include tasks such as weeding and mulching. Some volunteers assist scientists with research. They make sure plants are clearly labeled. They collect the plants for study. Karen Kerkhoff works at a botanical garden in Overland Park, Kansas. She said, "The most fun for me is when I see a family has come in. The kids may find a good stick walking the trail, and life is good."[2]

Those interested in volunteering at a botanical garden may need to submit an application. Volunteers may be assigned different roles based on how much gardening experience they have. Additional training may be necessary before volunteers can interact with plants and visitors. But even those with limited experience can find this volunteer work rewarding.

BUILDING A COMMUNITY GARDEN

Community gardens provide other volunteer opportunities. A community garden is a shared plot of land that a group of people tends. They can use the land to grow fruits, vegetables, and herbs. These gardens let city dwellers have access to green spaces.

Many cities and towns have community gardens. The American Community

Community gardens need volunteers to help start and operate them.

Gardening Association website has a list of these gardens throughout the United States. Interested gardeners should contact the person in charge of the community garden they want to join. An individual can join the garden. A group of friends and family members can also join together.

A community might not have a shared garden yet. People can work together to start a new community garden. They need to find an area of land for the garden. They also need approval from the local government to use the land for gardening. They develop rules for their gardeners to follow, such as not picking fruits and vegetables that belong to others. They may ban the use of **pesticides**.

Creating a community garden helps the environment. People can grow healthful foods. They do not need to depend on faraway farms for fruits and vegetables.

Crops from large farms often need to be transported long distances. Transportation releases carbon dioxide and other gases that cause climate change.

Community gardens also encourage healthful diets. They teach people to work together. Research has shown that gardening improves emotional health. It can reduce symptoms of anxiety and depression.

URBAN FARMING

Some people confuse community gardening with urban farming. But urban farms are more like traditional farms. They are businesses. The crops are grown to be sold to local markets.

CHAPTER TWO

HOW CAN I VOLUNTEER AT A CLEANUP?

People can volunteer to help the environment by taking part in a cleanup. There are many cleanups in the United States and worldwide. People can choose to join a nearby cleanup. They can also choose to organize their own.

Most local parks need volunteers to help pick up trash and keep the parks clean.

Volunteers collect trash at beaches and rivers. Some pick up trash along roads. Then they get rid of the waste safely. Picking up trash protects wildlife. It keeps natural spaces clean for people to enjoy.

Cleanups have visible results. This can make volunteers feel more connected to the environment. Nicole Wright, an environmental studies major at Connecticut College, volunteered at a beach cleanup. She said, "Recycling, taking public transportation, etc., are valuable ways to make a difference, but you can't always see the direct impact they have. Going to a beach and picking up trash is a way to make a direct physical difference, and it definitely feels great to see the impact we make."[3]

GETTING INVOLVED

People who are interested in joining a nearby cleanup can locate one online. Organizations such as Earth Day Network and American Rivers have websites to help volunteers find cleanups. Cleanup leaders often provide materials such as trash bags.

REDUCE, REUSE, RECYCLE

People can help the environment by lowering the amount of waste they produce. They can do this by remembering to reduce, reuse, and recycle. A student can reduce waste in the cafeteria by taking only what she needs. She doesn't grab extra napkins. She might reuse school supplies from an earlier year. She recycles papers she no longer needs.

Those organizing the event make sure to properly dispose of collected waste.

But there are things volunteers should know to prepare for the cleanup. They should wear closed-toe shoes and clothes they don't mind getting dirty. They can protect themselves from harmful sunlight by wearing sunscreen. Gloves should also be worn when picking up trash. Gardening gloves that can be reused are a good choice.

Volunteers should also know how to collect and dispose of waste. Recyclable items should be separated from trash.

Using gardening gloves instead of disposable gloves helps cut down on trash produced by a cleanup crew.

Many plastic, glass, and metal items can be recycled. Some items need to be disposed of in a special way. This includes items such as medical needles, batteries, and electronics. Needles should be thrown away in medical waste containers. Cities may have different rules on how to dispose of batteries and electronics. Often these items can be recycled at specialized recycling plants. Broken glass should not be thrown away in trash bags. It could cause the bags to tear and trash to spill out.

LIVING PLASTIC FREE

Plastic Free July is a global movement. It encourages people to stop using single-use plastics for a month. Single-use plastics include things such as straws and coffee cups. Many foods and supplies are packaged in plastics. The World Wildlife Fund offers tips to reduce plastic use. It suggests using reusable bottles and bags. Using bar soap instead of bottled soap also cuts down plastic use.

ORGANIZING A CLEANUP

In addition to joining a cleanup, people can organize their own. Cleanups can be big or small. Those organizing a cleanup need to have a clear plan of what they will do before, during, and after the cleanup.

The organizer first decides on a location. He buys materials, such as trash bags, for the cleanup. He may buy gloves for the volunteers or ask them to bring their own. He should also have hand sanitizer so volunteers can clean their hands. He should think about how volunteers will sort the waste they collect. For example, he

Participating in a cleanup can be a way to spend time with family or friends outdoors.

should have separate bins for trash and recyclables. He may have a container that can be sealed for broken glass.

The organizer should clearly communicate with volunteers when the cleanup begins. He should remind volunteers not to pick up items that are unsafe. Needles and syringes need to be handled carefully. Volunteers should not pick these up if they feel uncomfortable. He should also have a plan if a volunteer feels sick or gets injured while volunteering.

Waste needs to be disposed of properly after the cleanup. The organizer may plan to take trash to the landfill by himself. He could also schedule a garbage truck to arrive after the cleanup. He makes sure that volunteers are able to clean their hands and arms. He thanks volunteers for their service. He might share photos of the cleanup online. This may inspire others to organize cleanups in the future.

CLEANUP BENEFITS

Cleanups near rivers and lakes are especially important. Waste can contaminate waterways. This means the water is dirty and unsafe. Drinking this water can cause sickness. People can no longer use the water for swimming or fishing.

Cleanups are also important for wildlife. Some types of trash take a long time to **biodegrade**. Materials such as plastics take especially long to break down. Animals might eat trash items by mistake. This can

Clean beaches are healthier for the environment and more enjoyable to visitors.

cause choking or other injuries. The animal may die as a result. When plastic does break down, it releases **toxic** chemicals. These chemicals can make animals sick.

Volunteering at a cleanup has many benefits for the environment. It keeps natural spaces such as beaches healthy.

Animals in the area are protected from injuries caused by human waste. Cleanups also have community benefits. They bring people together to work toward a common cause. Afroz Shah is a lawyer in India. He also promotes beach cleanups. He said, "Volunteers who come to pick up are also getting trained to handle plastic. Anybody who sees plastic here will not buy plastic later."[4]

People are able to safely explore trash-free areas. For example, they do not need to worry about stepping on broken glass at a clean beach. Clean water is safe

HOW LONG TO BIODEGRADE?

Styrofoam
unknown; possibly never

Aluminum cans
80–200 years

Batteries
100 years

Nylon fabric
30–40 years

Plywood
1–3 years

Source: Rick LeBlanc, "The Decomposition of Waste in Landfills," *Balance Small Business*, January 16, 2021. www.thebalancesmb.com.

Some materials take a much longer time to biodegrade. It is important to avoid using items that take a very long time to biodegrade. That way they do not end up in nature.

for human activities. Clean environments

encourage people to spend more

time outdoors.

CHAPTER THREE

HOW CAN I VOLUNTEER AT A NATIONAL PARK?

There are sixty-three US sites with *national park* in their names. A national park is an area of land protected by the US government. National parks have beautiful landscapes and wildlife. Scientists study the animals and plants in the parks.

Employees and volunteers are needed to keep national parks running smoothly and to protect the parks' natural beauty.

These parks attract many visitors. The area of land in a national park is protected from pollution. People work to keep the park clean. National parks have natural resources such as trees and water. The resources are protected from human use.

Many national and state parks have volunteer opportunities. Duties vary depending on age and experience. Some people greet visitors at park centers. Others help repair trails. There are many ways to help. Volunteering at a national park is a great option for those passionate about the environment.

NATIONAL PARK VOLUNTEER OPPORTUNITIES

Volunteers at national parks are of many ages and backgrounds. Some positions require previous experience. Others are open to anyone.

Some national park volunteers help answer visitors' questions.

A campground host keeps the campground area clean. She provides information about the area to campers. She should have camping experience. Additional training may be needed so she knows her way around the campground.

Other opportunities also require special skills. These include serving as a trail guide or assisting with education. Educational volunteers may lead activities that teach others about the environment.

Some volunteers work with wildlife. They may help scientists with animal tracking. They may study how animal populations change over time. Others work to control invasive plant growth in the area.

Teens can sometimes volunteer in visitor services. These volunteers provide visitors with information about the park. They may hand out maps and brochures. They make

Volunteers may help remove invasive plant species at a national park.

sure visitors stay safe and have a good time at the park. Teens are also able to volunteer to help record information at national parks. They may collect data on how many people visit certain trails in a park.

Many national park volunteers want others to experience the beauty of nature. Tom Crochetiere is a volunteer at Joshua Tree National Park. He was asked to describe his work as a volunteer. He said, "I am thankful for the opportunity to help educate visitors from all over the world about our park. . . . I am empowering them to love our park much in a way that we all do."[5]

GETTING INVOLVED

People interested in volunteering at a national park need to submit an application to the National Park Service (NPS).

Keeping trails clear for visitors is an important task. Volunteers can help with this.

Descriptions of available positions can be found at www.volunteer.gov/s. Applicants under the age of eighteen need permission from a parent or guardian. An applicant should know what days of the week she can help.

The volunteer also notes the positions that interest her. She may apply for work with wildlife or as a tour guide. She notes any skills she has that would help her as a volunteer. She may have experience with public speaking. She could know first aid. These things help volunteer managers match volunteers to a national park.

If she is a good match for a position, a volunteer manager will set up an interview. The applicant discusses why she wants to be a volunteer. Sometimes there are no available positions. But positions may open later.

The NPS also has programs with Girl Scouts and Scouts BSA. Scouts can work with park leaders to do service projects. They learn about the environment and cultural history of a park. Park rangers may teach scouts about protecting the environment. They help scouts learn

YOUTH CONSERVATION CORPS

Some people are passionate about the outdoors. The Youth Conservation Corps (YCC) is a good option for those interested in one day helping the environment as a career. The YCC is a summer program for teens ages fifteen to eighteen. They are paid to work in national forests and parks. For forty hours a week, teens in the YCC repair trails and help with wildlife research.

about the park through volunteering. Scouts may earn a certificate or patch for their involvement.

Scouts can work on trail repairs. They may help build a footbridge. They might replant trees after a wildfire. Older scouts can do long-term service projects on their own. A scout may clean up an area of the park. He may build and install signs about fire safety at campsites. He can create educational videos for the park. Videos may show how to take care of the environment. These projects may help a scout earn high honors.

BENEFITS OF VOLUNTEERING

Volunteers help make sure that visitors enjoy their time at national parks. They keep parks safe and clean. Their work helps visitors have positive experiences with nature.

GEORGE AND HELEN HARTZOG AWARDS

The NPS recognizes some volunteers for their outstanding service. In 2017 Nicholas Gilson won the Hartzog Youth Award for his Eagle Scout project. A fence had broken along the Ice Age National Scenic Trail in Wisconsin. As a result, hikers were harming native plants along the trail. Nicholas raised more than $1,200 to repair the fence and build benches by the trail. His project protected local wildlife. It also improved visitors' experiences of the trail.

Some volunteers teach school groups at national parks.

Volunteers and visitors may feel encouraged to spend more time outdoors.

Spending time in nature has many health benefits. It improves mental health. Being in nature reduces stress. It also lowers levels of anxiety and depression. Dr. Jason Strauss is a director of **psychiatry** for the elderly at Cambridge Health Alliance.

He said, "Having something pleasant to focus on like trees and greenery helps distract your mind from negative thinking."[6] Walking outside can help children focus at school. Memory and concentration improve from being outdoors.

Outdoor activity also improves physical health. It can lower the risk of catching certain diseases. Physical activity makes the heart healthier. The work that volunteers do encourages people to spend more time outdoors. National park visitors and volunteers can live healthier and happier lives.

CHAPTER FOUR

HOW CAN I VOLUNTEER AS AN ACTIVIST?

National governments make laws to protect the environment. Many organizations help raise awareness about environmental issues. But young people can also make their voices heard.

Greta Thunberg is an environmental activist. She is from Sweden. Thunberg was

Greta Thunberg participates in a demonstration about climate change. Her sign in Swedish translates to "School Strike for Climate."

named Time magazine's Person of the Year in 2019. She was sixteen years old at the time. In August 2018, she began skipping school to protest outside a Swedish government building. She held a sign that read, "School **Strike** for Climate." These actions attracted the attention of world leaders. She met with some of them. Thunberg urged leaders to take concerns about the environment seriously. She said, "We can't just continue living as if there was no tomorrow, because there is a tomorrow."[7]

Thunberg's actions show that young people can make a difference for the environment. Thousands of people joined her protests in Sweden. Other students around the world followed her example. They led school strikes for the climate.

There are other ways for young people to raise awareness about climate change. They can stay informed about laws and policies that affect the environment. They may start a school club. They can support organizations that focus on the environment.

SUPPORTING LOCAL BUSINESSES

Shopping at local businesses is an eco-friendly choice. Small, local businesses generally take less energy to run than large businesses. Items bought from local stores do not have to be shipped long distances. Locals can come pick up their orders. This reduces waste from transportation and packaging.

PARTICIPATING IN GOVERNMENT

Governments can make laws to protect the environment. These laws may limit the use of fossil fuels, which hurt the environment. Fossil fuels include oil, natural gas, and coal. Companies may have to pay a tax if they use large amounts of these materials. This may influence them to reduce their usage. Governments may invest in clean forms of energy such as solar and wind energy. These steps would lower **emissions**.

　　Laws can make a big difference for the environment. People who are able to vote can elect leaders whose ideas

Burning fossil fuels, such as the gasoline used in cars, contributes to climate change.

match their own. Children and people who are not able to vote can still write to government leaders. They can explain why environmental issues are important to them. They can make sure that their voices are heard.

For example, Mari Copeny gained national attention for her letter to President Barack Obama in 2016. She urged him to address environmental concerns. At the time, her community in Flint, Michigan, was facing a water crisis. The water supply had high levels of lead. Drinking this water could cause severe health problems and even death. Obama received her letter and visited the city. He listened to residents of Flint and called for a government response to the crisis.

Young people can also attend strikes about the environment. Many people come

Members of the National Guard distributed water bottles throughout Flint so people could have drinking water.

together to support a common cause during a strike. In September 2019 more than 4 million people protested climate change worldwide. They carried signs and called upon governments to make changes. These demonstrations can attract the attention of lawmakers. Politicians can work to create policies that address the environment. Children and teens can also sign petitions to show their support for certain policies.

STARTING A CLUB

Starting a school club is another way to spread awareness about the environment. Interested students should come together. They should find a teacher who is willing to lead the club. The students and teacher think about projects the club can do.

Students around the world hold strikes to encourage their governments to address climate change.

Before the first meeting, students make sure that others know about the new club. They talk to their friends about coming to the meeting. They may hang posters about the club in the school hallways. They ask

A school garden can be a place to learn as well as a source of locally grown food.

other teachers to share club information with their students.

The club brainstorms ideas about projects that help the environment.

Members set goals for what they want to accomplish. This helps the club stay focused. Projects should involve all club members. They should be realistic. The club should be able to accomplish its goals within a school year. Organizing a cleanup or starting a school garden are two choices. The club could organize an environmental awareness day to teach classmates about the environment and ways to protect it. Members may explore the outdoors together. The teacher may plan a trip to a state park for students to learn more about nature.

The club should continue to look for new members. Members should welcome anybody who wants to help out. They may reach out to community members for fundraising and donations. For example, a club could plan to start a school garden. Members may reach out to the community for shovels and other garden tools.

Students and teachers can take pictures of the project. Pictures help them see how the project has developed over time. This can motivate students to keep up their hard work. It may help inspire new students to join the club. Bigger clubs can

School nature trips can be a great way to learn more about the environment.

tackle bigger projects. Once the project is complete, it is time to start a new one.

SUPPORTING ENVIRONMENTAL ORGANIZATIONS

Many organizations focus on the environment. These organizations provide ideas about how to get involved. They provide resources that explain important environmental issues. Staying up-to-date on information about the climate is important for young activists. They influence government leaders to make laws that protect the environment. They support scientists who fight for environmental health.

Environmental organizations call people to volunteer. They put together events such as cleanups or strikes. They provide information about the environment. Donating to these organizations is another

CARBON FOOTPRINTS

Households can help the environment by managing their carbon footprints. A carbon footprint describes how much a person impacts the environment. An eco-friendly person has a small carbon footprint. She makes choices to not hurt the environment. There are many tools online to calculate a household's carbon footprint. They use information such as how much energy a family uses to heat the home. They ask about the family's transportation. Carpooling and recycling are two ways to reduce a household's carbon footprint.

Local hubs of the Sunrise Movement can help keep students informed about ways to get involved, such as attending protests.

way to show support. Money can be used to further scientific research.

The Sunrise Movement is a youth-led environmental organization. Its goal is to stop climate change. It pushes for new jobs to be created in clean energy. It encourages people to talk to community members and leaders about the environment.

Members of the Sunrise Movement texted and called young voters during the 2020 US presidential election. They encouraged voters to elect candidates who would fight for the environment. When Joe Biden was elected as president, they urged him to choose leaders who would work to stop climate change. Varshini Prakash is a cofounder of the Sunrise Movement. She said, "This moment not only calls upon us to call our politicians

People can encourage others to sign petitions to help the environment.

to action, but for all of us to recognize the [ability] that we have . . . to push forward on this issue."[8]

Members created an online petition to show the importance of environmental issues. People could sign the petition. People can also talk to friends and family about the environment. These are just two ways to show support of environmental laws.

From gardening to cleanups, people of all ages can make a difference. Volunteering outdoors has many benefits for the community and for oneself. Taking time to care for the planet makes the world a cleaner place for the future.

GLOSSARY

biodegrade
to break down into smaller parts by natural processes

emissions
gases that are put into the environment

invasive species
types of plants or animals that have come to an area where they do not naturally live and are hurting native species in the area

pesticides
chemicals that are used to kill pests such as bugs

psychiatry
the study of mental, emotional, and behavioral health

strike
a form of protest characterized by the stopping of work or school attendance; used to raise awareness of an issue

toxic
poisonous or harmful

urban
of or relating to a city

SOURCE NOTES

CHAPTER ONE: HOW CAN I VOLUNTEER AT A GARDEN?

1. Quoted in Anna Liz Nichols, "Community Gardens Could Lead to Lower Crime Rates in Cities," *State News*, April 4, 2019. https://statenews.com.

2. Quoted in Debra Skodack, "In Tough Times, Overland Park Arboretum & Botanical Gardens 'Beneficial to the Soul,'" *Kansas City Star*, September 8, 2020. www.kansascity.com.

CHAPTER TWO: HOW CAN I VOLUNTEER AT A CLEANUP?

3. Quoted in "First-Year Students, Staff Adviser Help Clean Up Ocean Beach," *Connecticut College*, September 23, 2020. www.conncoll.edu.

4. Quoted in Kathleen Toner, "He's Doing the 'Dirty Work' to Keep Plastic out of the Ocean," *CNN*, October 17, 2019. www.cnn.com.

CHAPTER THREE: HOW CAN I VOLUNTEER AT A NATIONAL PARK?

5. Quoted in "Meet Our Volunteers: Tom Crochetiere," *National Park Service*, April 8, 2019. www.nps.gov.

6. Quoted in "Sour Mood Getting You Down? Get Back to Nature," *Harvard Health Publishing*, July 2018. www.health.harvard.edu.

CHAPTER FOUR: HOW CAN I VOLUNTEER AS AN ACTIVIST?

7. Quoted in Charlotte Alter, Suyin Haynes, and Justin Worland, "TIME 2019 Person of the Year: Greta Thunberg," *TIME*, n.d. https://time.com.

8. Quoted in Tonya Mosley and Allison Hagan, "Sunrise Movement Climate Activists Call On Biden to 'Keep His Promises,'" *WBUR*, November 20, 2020. www.wbur.org.

FOR FURTHER RESEARCH

BOOKS

Dan Hooke, *Climate Emergency Atlas: What's Happening—What We Can Do*. London, United Kingdom: DK, 2020.

Walt K. Moon, *Volunteering for Animal Welfare*. San Diego, CA: BrightPoint, 2022.

INTERNET SOURCES

"The History of Earth Day," *Earth Day Network*, n.d. www.earthday.org.

"Volunteer with Us," *NPS*, September 2, 2020. www.nps.gov.

WEBSITES

The Sunrise Movement
www.sunrisemovement.org/about

The Sunrise Movement is a youth-led organization. It unites young people to stop climate change.

Volunteer.gov
www.volunteer.gov/s

Volunteer.gov lists opportunities for volunteering at national parks and other national sites.

RELATED ORGANIZATIONS

Earth Day Network
1752 N Street NW, Suite 700
Washington, DC 20036
info@earthday.org
www.earthday.org

Earth Day Network is a worldwide organization that strives to educate and encourage environmental activism. It calls for individuals to make their voices heard in their communities and governments.

National Park Service
1849 C Street NW
Washington, DC 20240
www.nps.gov

The National Park Service oversees and maintains US national parks. The NPS works with volunteers and local governments to celebrate natural and historic spaces throughout the United States.

Ocean Conservancy
1300 19th Street, NW, 8th Floor
Washington, DC 20036
memberservices@oceanconservancy.org
https://oceanconservancy.org

The Ocean Conservancy is an organization dedicated to ocean health. It works with scientists and volunteers to protect the ocean and its wildlife.

INDEX

American Community Gardening Association, 22–23
American Rivers, 29
Arbor Day, 6–7

beaches, 27–28, 37–38
Biden, Joe, 71
biodegrading, 36, 39
botanical gardens, 13, 18–22

campground host, 43
carbon footprints, 69
city parks, 13–18
cleanups, 11, 26–39, 65, 69, 73
community gardens, 13, 22–25
compost, 15
Copeny, Mari, 60
Crochetiere, Tom, 46

Earth Day Network, 29

fossil fuels, 58

Gilson, Nicholas, 51

Hartzog Youth Award, 51

Kerkhoff, Karen, 21

National Park Service (NPS), 46, 49, 51
national parks, 40–53
natural resources, 41

Obama, Barack, 60

petitions, 62, 72–73
Plastic Free July, 32
Prakash, Varshini, 71
protests, 55–56, 61

reduce, reuse, recycle, 29

Sadler, Richard, 18
school clubs, 57, 62–68
scouts, 49–50, 51
Shah, Afroz, 38
shopping locally, 57
Strauss, Jason, 52
strikes, 56, 60–61, 69
Sunrise Movement, 70–71

Thunberg, Greta, 54–56

urban farming, 25

visitor services, 44

wildlife, 27, 36, 40, 44, 48, 49, 51
Wright, Nicole, 28

Youth Conservation Corps (YCC), 49

IMAGE CREDITS

Cover: © Prostock-studio/Shutterstock Images
5: © bodnar.photo/Shutterstock Images
7: © DUO Studio/Shutterstock Images
8: © wavebreakmedia/Shutterstock Images
10: © Sergei Domashenko/Shutterstock Images
13: © refrina/Shutterstock Images
17: © JJFarq/Shutterstock Images
19: © Wileydoc/Shutterstock Images
20: © Jose Maria Ruiz Sanchez/Shutterstock Images
23: © Monkey Business Images/Shutterstock Images
27: © Africa Studio/Shutterstock Images
31: © Inside Creative House/Shutterstock Images
34: © David Pereiras/Shutterstock Images
37: © Deborah Kolb/Shutterstock Images
39 (top left): © Ashusha/Shutterstock Images
39 (top right): © Denis Semenchenko/Shutterstock Images
39 (bottom left): © MarySan/Shutterstock Images
39 (bottom middle): © Belle Hung/Shutterstock Images
39 (bottom right): © NiRain/Shutterstock Images
41: © Galyna Andrushko/Shutterstock Images
43: © Sundry Photography/Shutterstock Images
45: National Park Service
47: National Park Service
52: National Park Service
55: © Liv Oeian/Shutterstock Images
59: © Paolo Bona/Shutterstock Images
61: © Linda Parton/Shutterstock Images
63: © John Gomez/Shutterstock Images
64: © DeepGreen/Shutterstock Images
67: © Maciej Dubel/Shutterstock Images
70: © Sundry Photography/Shutterstock Images
72: © Monkey Business Images/Shutterstock Images

ABOUT THE AUTHOR

Chelsea Xie is a writer who lives in San Diego, California. She enjoys reading poetry, cross-stitching, and spending time outdoors.